DIAMOND: ETERNAL FIRE

Diamond: Eternal Fire

The Indian Diamond Tradition

AshaRani Mathur

Rupa & Co

ETERNAL INDIA

Published 2006 by
Rupa & Co
7/16, Ansari Road, Daryaganj
New Delhi 110 002

Sales Centres:
Allahabad Bangalore Chandigarh Chennai
Hyderabad Jaipur Kathmandu
Kolkata Mumbai Pune

Book Design & Typeset by
Arrt Creations
45 Nehru Apts Kalkaji, New Delhi 110 019
arrt@vsnl.com

Printed in India by
Gopsons Papers Ltd., Noida.

Page 1: Pendant, diamonds and coloured stones, from the Nuvo Collection
Page 2: Bracelet, diamonds in gold, detail from a South Indian design
Page 3: Pendant, diamonds and emeralds, Jaipur
Page 5: Nose ornament, diamond beads and set diamond with emerald, Rajasthan/Gujarat

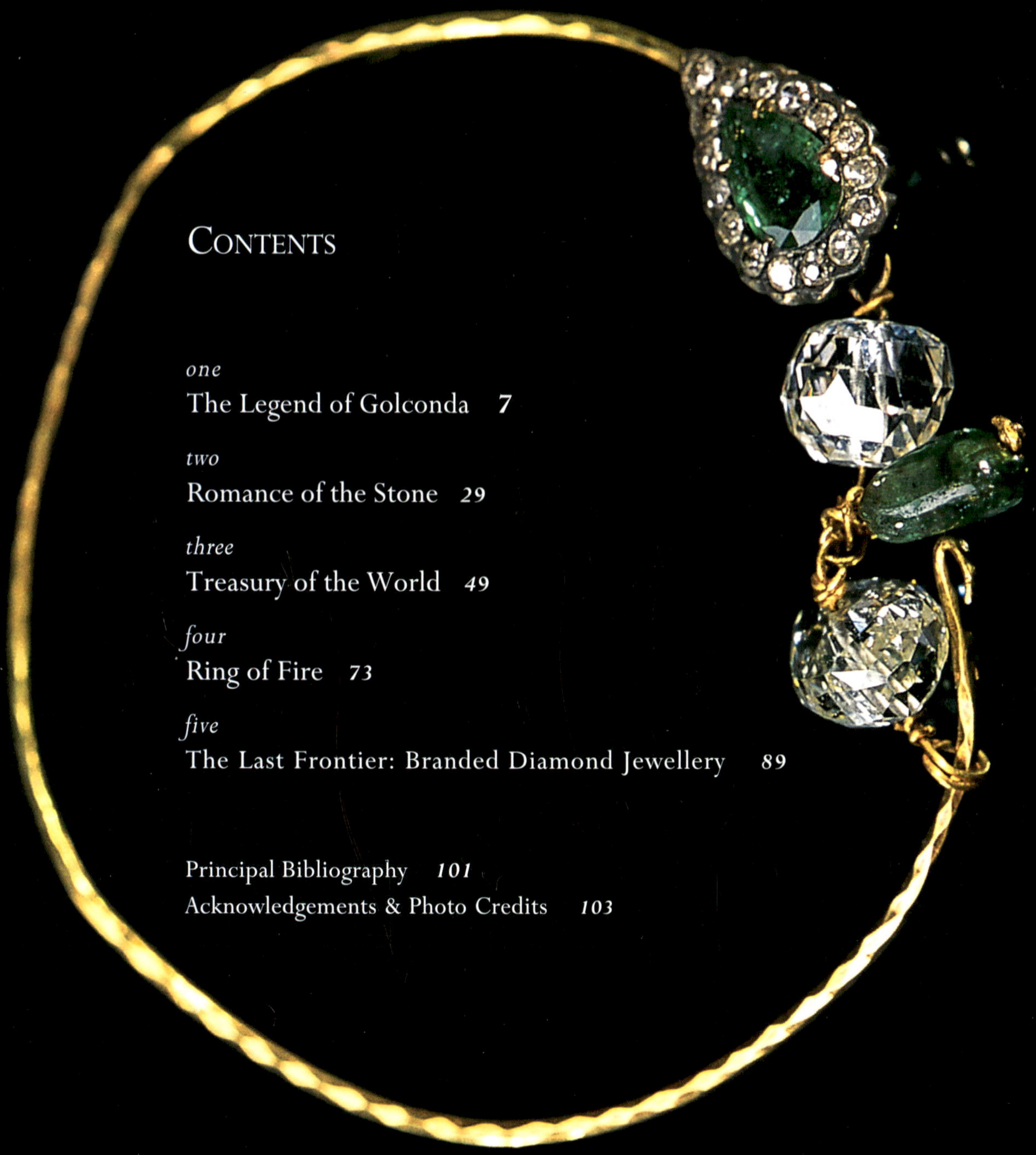

Contents

one

The Legend of Golconda

"It has... diamonds such
they have no price or number..."

"Goa", *Garcia de Resende*

For centuries India was the greatest and only source of diamonds known to the world. True, there were mines in Borneo, but they could not compare with the sheer output and quality of the gems from India; from about 800 BC to well into the eighteenth century, the stones poured out of Indian mines, their sources jealously guarded by local rulers who demanded that the finest be kept for them. A trade that traced back to ancient India and the empire of Rome

opp. Bajuband, Karnataka, early 19th cent.
Armlet in gold, rubies and diamonds. The lustre of the diamonds indicates their source as Golconda.

Tikka, Mughal, 18th cent.
This Mughal period head ornament with a floral motif has foil-backed table cut diamonds and emeralds in the outer rim.

was sustained by an army of miners, middlemen and merchants, and a route that in medieval times extended from the west coast of India to Venice and beyond.

It was not surprising, then, that when at last the Europeans arrived in India their prime pursuit was the legendary wealth of the country, its gems and spices, its textiles and gold. Vasco de Gama, the Portuguese admiral, made landfall off the coast of Malabar on a warm and muggy summer's day in 1498. One of the first encounters his men had was with a Tunisian merchant who exclaimed: "A lucky venture, a lucky venture! Plenty of rubies, plenty of emeralds! You should thank God for having brought you to so rich a country!"

This mention of gemstones could well have included diamonds. To the east of the coast lay the fabled empire of Vijayanagara, and Golconda in the Bahmani kingdom of the Deccan, each bursting with gemstones of unimaginable variety and richness. But within decades the export trade in precious stones was wrested from the Arabs and taken over by the Portuguese through their prime colony of Goa: Goa, which rapidly became the greatest Asian trading point for diamonds, rubies, sapphires, pearls and topazes, amongst other stones. There was a massive exodus of gems and jewellery from India, much of it passing through the experienced hands of the lapidaries and craftsmen of the Rua Direita in Goa, the street of jewellery ateliers.

Ear studs, Jaipur
Diamonds cut in parab style are close set to form flower shape studs. The back of the ear ornament is delicately worked with a floral outline studded with diamonds.

There were many participants, one way or another, in the swelling export of gems. Jacques de Coutre settled in Goa in the early seventeenth century; for twenty years thereafter, he ran a flourishing business over much of Asia, trawling the mines of the Deccan for diamonds. In the mid-eighteenth century, when diamond production in India was already on the decline, the Portuguese viceroy, the Marquis of Alorna, returned to Europe with a rich haul of diamonds, chestfuls amounting to an amazing 4,500 carats.

Bracelet, Mughal, 18th cent. (Benaras)
Note the Golconda diamonds, cut so as to enhance their size and value; such cutting was a speciality of Benaras, as was the pink enamelling seen on the back.

Though gems and precious metals had been traded through the ages, we can see here the beginnings of that great dispersion of Indian riches into Europe that has left us today, in the twenty-first century, with a meagre portion of our heritage. Traditionally, the locations of gemstone mines

were kept secret by the local rajas who collected duties based on the value of their outputs. Moreover, they held the right to keep the largest stones for themselves; indeed, in many parts, stones whose weight exceeded a certain number of carats could not be sold elsewhere.

Bangle, Deccan setting and workmanship, 19th cent
The single row of diamonds is backed by a lotus motif in pink enamelling; perhaps the enamellists came from Benaras, of which this colour is characteristic.

Bracelet, Deccan, early 18th cent.
These are almost certainly Golconda diamonds, foil-backed for brightness. Note the cut, which makes the diamonds appear larger. The clasp is in the makara or crocodile shape.

The result was vast accumulations of precious stones in imperial and royal vaults. Such a treasure had many uses. First and foremost was the power that this enormous wealth concentrated in the hands of the rulers, a strength augmented by the mystical and magic properties believed to be inherent in the gems and thus transferred to their owners. The gems could be fashioned into sumptuous jewellery worn on state occasions or could be embedded in thrones, the grandeur and opulence of perfectly-matched ruby drops and brilliant diamonds a dazzling affirmation of sovereignty to intimidate friends and foes alike. They could be worked into elaborate offerings for the gods and presented to temples, a thanksgiving for Divine protection or victory in war, a fervent prayer for its continuance. Gems were easily portable, as more than one monarch fleeing his enemies found out, and could be just as easily converted into money to pay for armies. Many a long and costly military campaign was financed by gems, versatile as payment for information, the purchase of loyalty or an inducement for defection. And such expenditure could be recouped by the sacking and pillaging of the treasuries of the defeated.

Thus, until at least the tenth century, the best gems remained in India; and even though the depredations of invading armies in medieval times lightened some treasuries, enough of the cache was hidden away or buried to serve as the focus for future collections.

The gemstone that aroused the greatest awe was the diamond whose hardness and brilliance made it the most sought-after and expensive. And for centuries it was the name of Golconda which was closely associated with the stone, a name that was a synonym for large diamonds of the

Necklace, Rajasthan, early 20th cent.
The choker style necklace is adorned with good quality diamonds in kundan setting. The diamonds are graded from the larger one at the centre to the smaller ones at the sides.

utmost purity and clarity, whiter than white, whose lustre and transparency brought together the luminosity of water with the flash of fire. So desirable were these qualities that till today, long after the mines lie exhausted, the most rare and valued stones for collectors around the world remain identified with Golconda.

The most glorious diamonds in all of India were traded in the vicinity of the Golconda fort in the bazaar of Caravan-i-sahu. Here came stones from the twenty-three diamond mines of the kingdom of Golconda, mines which had been in existence for long before the Qutb Shahi dynasty came to power there in AD 1518. Here too came diamonds from surrounding areas; all these superb stones were generically known as Golconda diamonds. Golconda fort itself was built on the ruins of the mud fortess of the Rajas of Warangal. In the early fourteenth century the armies of Alauddin Khilji defeated Warangal, sacked the fortress and departed with the booty which was colossal. One amazed contemporary writer described it thus: "The boxes carried by the elephants were full of valuables and gems, the excellence of which drove the onlookers mad...The diamonds were of such a colour that the sun will have to stare hard for ages before the like of them is made in the factories of the

opp. Bajuband, Tamil Nadu, early 19th cent.
Diamonds and rubies set in gold with an emerald at the centre.
The gold has the weighty appearance characteristic of South India.

rocks…Every diamond sparkled brightly like drops fallen from the sun…Among them was a jewel unparalleled in the world".

These "unparalleled jewels" came from sources scattered over the areas on or between the Pennar and Krishna rivers in present day Andhra Pradesh and Karnataka. There were three groups of mines, the Cuddapah on the Pennar river, the Nandiyal between the Pennar and the Krishna, and the Elluru or Golconda on the lower Krishna. Among these were the famous mines of Kollur, Partiyala, Kondapettah, Banganapalli, Ramulkota and Malavelli: names recorded in history as the birthplaces of some of the largest and most renowned diamonds in the world.

Sarpech, Rajasthan, 20th cent.
Diamonds surround a semi-precious stone in this turban ornament, and drop pearls complete the design.

And yet, it is almost incongruous to consider the humble beginnings of these stones mined from alluvial deposits. Diamond mining in India was always an intensely laborious process; in the kingdom of Golconda, as elsewhere, its workers received paltry wages that bore no relation to the huge sums that changed hands in the trading bazaars. Guards kept a watchful eye to prevent theft on site, though not always successfully, for practiced miners had ingenious ways of smuggling out the stones.

The methods used to excavate the gemstones varied from one region to the other depending on how close to the surface the diamond veins were. For descriptions of these methods we have to thank Jean-Baptiste Tavernier, the seventeenth century traveller and connoisseur of gems, who visited India and kept meticulous records of his observations. In Kollur, for example, the earth was excavated to the depth of the water table and then washed in previously erected enclosures with exits that allowed for the outflow of water and slime. Such washing might be done two or three times, and the resultant soupy clay left to dry through the heat of the sun into a coarse caked residue. This was then winnowed so that the fine dust flew out. What was left was spread out and pounded with wooden pestles, winnowed again, and searched for diamonds.

In Ramulkota, as also in Banganapalli, the method followed was different. Here, iron picks with curved ends were thrust into diamond veins along the direction of the vein to access the purest and most colourless stones. The sand or earth drawn out was then put into vessels and washed in search of the stones. In some areas, the stones were mined at great depth, but in Kondapettah, for example, the cut could be as little as four feet deep, from which was extracted pebbles and gravel. These were washed in specially constructed reservoirs to locate the diamonds, which were "easily recognisable in the moist state by their peculiar lustre".

Baksua, buckle, Deccan, early 19th cent.
This piece, with its spinels and Golconda diamonds, is probably from the Nizam's collection. The centre emerald is in open setting, the diamonds in closed setting.

Mining the diamonds, however, was only the beginning of the pipe-line that led from production to sale, part of a highly organised yet tightly controlled venture. The mines were given out on lease, earning the Golconda exchequer a sizeable revenue. Diamond merchants purchased from the mine manager in lots, usually smaller stones of varying value, bought against cash or bills of exchange. Then, as now, the diamond trade was in the hands of Gujarati merchants who had gradually gained a monopoly over this lucrative business. There was trading in

Karnaphool, ear studs, Jaipur, 19th cent.
Indian diamonds embellish this magnificent pair where the stones are in kundan setting. Note the difference in cutting between the petals (flat cut) and the centre diamond, which has more facets.

the marketplace too, and Tavernier noted "...the seller and buyer talk neither by means of their lips nor their eyes, but only by the hand..." In this silent and arcane system of hidden signals, the seller and buyer faced each other, and the seller negotiated his price by taking the right hand of the buyer, which he carefully concealed under his waistband. The whole hand stood for a certain sum, five fingers indicated another sum, half a finger yet another and so on. In this manner the deal was struck, money changed hands, and a single parcel of diamonds could be bought and sold five or six times over without anyone being the wiser as to its selling price in successive transactions. As the weight of each lot was stated by a special representative of the King, there was little risk of deception.

The largest diamonds were reserved for royalty; in Golconda no stone over ten carats in weight could be sold as it was regarded as the King's property and had to be turned in to him. In later centuries, the successors of the Qutb Shahi kings of Golconda were the Nizams of

Hyderabad, an image we associate today with wealth and jewellery of fabulous proportions.

"...JEWELS MUST BE EVALUATED BY EXPERTS IN ORDER TO READ THE DESTINY THAT IS WRITTEN WITHIN THEM..."

FROM THE BRHATSAMHITA *OF VARAHAMIHIRA*

opp. Necklace, Jaipur
Designed in the Mughal style, this necklace of Golconda diamonds in polki cut is embellished with pearl drops and an upper fringe of pearls, and enamelled on the reverse.

The diamonds of Golconda were truly stones of destiny. Their unparalleled beauty lay in their spectacular size, their brilliance and incomparable whiteness. The list includes the best-known diamonds in the world, including the sublime Koh-i-noor, the Pitt-Regent, the Nassak, the Sancy, the Orlov, the Hope, the two Indore Pears, the Darya-i-noor, the Noor-ul-Ain, the Dresden Green, to name only a few. Alas, not one of these is now in India: what is more, re-cut and re-named, not only have they been sized down, but their identity as Golconda stones has been wiped out and so has much of their history.

There are differences of opinion about certain major diamonds. Was, for example, the Koh-i-noor the same stone as that known as Babur's Diamond, a

Bracelet, Jaipur
A triple row of diamonds forms the florals of this elegant bracelet, enamelled on the reverse.

Amulet, Deccani work in Mughal style, late 18th/early 19th cent.
The green on gold enamelling shows the Deccani origin. The Indian diamonds, perhaps from Golconda, are foil-backed for added brightness.

diamond so huge that its value was reckoned at two and a half day's food for the whole world? And was this, or either of these, the fabulous stone called the Great Mughal seen by Tavernier in Aurangzeb's collection whose original weight was an astounding 787 carats? Historians can speculate, but we may only marvel at the wealth implied by the stones and be saddened by their dispersal.

Such greatness exacted its own price in blood lust and greed; around the stones swirl tales of deceit, treachery and the tortuous path taken by each stone in its individual history. The deities of Hindu temples were routinely adorned with precious stones; in many cases, the eyes of the images were fashioned from diamonds. The Orlov diamond, "half the size of a hen's egg" and originally close to 195 carats, formed the eye of the image of Sriranga in the temple at Srirangam. From here it was stolen by a French soldier and eventually ended up as part of the Russian Imperial sceptre of Empress Catherine. Similar is the story of the Idol's Eye, said to be prised out of a sacred temple image, sold and re-sold over the many centuries of its existence.

Some diamonds were reputed for bringing bad luck to their owners; the Hope diamond, a magnificent violet-blue colour, was one such. It changed hands over and over; from being part of the French Crown jewels it came into the possession of a rich banker, Henry Hope, albeit in a re-cut and smaller form. But here too ill-fortune dogged the family whose last member went bankrupt. Now followed a sequence of murder, suicide, revolution and blood as ownership passed from one person to another. Even if part of the bizarre legends surrounding the stone is discounted, there are enough documented stories to warrant the aura of bad luck; the last owner, Mrs. Maclean, lost two children in tragic circumstances.

Of all the Golconda diamonds, surely the most gripping narrative is that of the Koh-i-noor, whose passage from hand to hand left a trail of blood. In 1739, when the Persian invader Nadir Shah laid waste the remnants of Mughal glory and stripped the treasury of all its jewels, he pillaged and killed his way through the capital of Delhi for close to two months. Among the treasures he sought was this diamond whose fame had spread through the world, but try as he might he could not find it. Finally the secret was betrayed: the diamond was hidden in the turban folds of the defeated Emperor Muhammed Shah. The trap was set; the wily Nadir Shah asked for an exchange of turbans as a gesture of friendship. Muhammed Shah had no option but to accept; and when the diamond was revealed in its full beauty, Nadir Shah exclaimed: Koh-i-noor! The Mountain of Light! And thus the name has remained since then.

But its travels did not end here. Murder, blinding and torture followed it from Persia to Afghanistan as kings and heirs fought over the stone. Eventually it returned to India with the fleeing Afghan, Shah Shuja, who sought shelter in Lahore, then under the rule of Maharaja Ranjit Singh of Punjab. The Maharaja offered to buy the magnificent diamond from Shah

opp. Necklace, Ariya, Rajasthan, late 19th cent.
This lavish choker with its pearl skirting has different sizes of diamonds studded in gold. Such a necklace would have been the possession of a wealthy patron, for "ariya" means "of the rich".

Shuja, who loftily declared: "If five stones were cast with strength, one to the north, one to the east, one to the south, one to the west and one straight up; if all the space between these points were filled with gold—that would still not equal the value of the Koh-i-noor". Nonetheless, the gemstone came into the hands of Ranjit Singh. After his death, and the annexation of Punjab by the British, it was taken by them as repayment for a "debt" and presented to Queen Victoria in 1850, who had it re-cut. Interestingly, she paid heed to the legend that said that the diamond would bring ill-luck to a man; thus, the brilliant stone now graces the crown worn by the Queen Consort of the British monarch.

In our times, we are used to associating the diamond with love and constancy: "A diamond is forever" goes the famous line, entwining the stone with the lover's gift of fidelity. The fact remains, however, that throughout history some of the most unspeakable acts have been committed to acquire diamonds, and that remains a tragic truth even today. A paradox. Just like those that exist within the mysterious heart of the diamond itself—the glitter of ice, the flash of fire, the pure sparkle of water, the very radiance of light.

opp. Necklace, Jaipur
Triple strand of Golconda diamonds set in 22 karat gold enamelled on the reverse.

two

Romance of the Stone

"'Mine,' said the god, 'must be the gem of the universe...' and lo! The diamond was born, pure as the dew drop and invincible in hardness..."

Paraphrased from

The Curious Lore of Precious Stones *by G.F. Kunz*

Ancient Hindus spoke of the creation of the diamond in mythological terms. There was once a demon named Vala, the text of the *Garuda Purana* tells us, whose growing power alarmed the gods. They wanted to destroy him, and lured him to his death on the pretext of a boon. He was "tied to a stake with thirteen strings" and slain; his body dismembered, and the severed limbs transformed into gem

Heart of Eternity
A rare and spectacular blue diamond sourced from the De Beers Premier mine in South Africa, it has an intense, saturate colour. It took more than three years to cut and polish the stone, which is the largest of the De Beers Millennium jewels displayed in the year 2000.

seeds. As the demigods jostled for the seeds, some fell to earth, scattering over the seas, rivers, mountains and forests. The seeds germinated into lodes of gemstones. Vala's blood was dropped by Surya the Sun God into deep pools where it transmuted into beautiful rubies; his teeth "fell like stars into the oceans below" and became lustrous pearls; his eyes turned into sapphires; in similar manner, from his other parts came emeralds, coral, topaz, zircon, cat's eyes... And from his calcified bones came the hardest gemstone of all—the diamond.

A compelling story, but the scientific reality is more prosaic, though equally fascinating. Imagine the entire weight of the Eiffel Tower, 7000 tonnes, resting on a five-inch plate; or 200 tonnes placed on a single coin; to this add temperatures exceeding 1000° C: that is the kind of tremendous pressure and heat needed to create a diamond. It is pressure of a sort that the forces of nature have sustained over a very, very long period of time, between one and three

above: Round brilliant cut (RBC) diamond with 57 facets, top view.

opp. Necklace, Jaipur
Rows of diamonds kundan set in soft gold form a large necklace with a pendant, almost like a breastplate. Pearl drops complement the diamonds.

Bracelet, Jaipur
The bracelet is enamelled on the reverse in a floral design in the typical red of Jaipur.

billion years. The stone that graces a crown, a necklace, a pair of earrings, has been witness to the very history of the earth, and has been present for at least two-thirds of its existence.

Briefly put, a diamond is the most concentrated form of carbon, formed deep within the earth's interior and brought to the surface through upheavals. Between 150 and 200 kilometres below the surface of the earth, under its skin-like crust and in the layer known as the mantle, geological processes crystallise the carbon at extreme pressures and very high temperatures. Diamonds are literally catapulted to the surface, carried by the rare molten rock, or magma, which erupts from very great depths in turbulent volcanic seizures. The channels which propel the diamond-bearing rocks are called kimberlite or lamproite pipes.

In India, the original mother rock of the diamond belonged to earlier geological periods, harking back to remote times. It rested in ancient terrain; its weathering over long periods of time caused it to manifest in conglomerates of sands and gravels washed by the waters of streams and rivers. It was in the alluvial deposits of river beds and their surroundings that the gemstone was primarily mined.

Bracelet, South Indian design
The gold and diamond bracelet has typically South Indian workmanship, as seen in design execution and setting.

The ancient Greeks gave the stone the name by which we know it today: diamond, from "adamas", translated variously as the conqueror, the invincible, the uncorrupted. There were many evocative descriptions of the gem, as frozen dewdrops or splinters from the stars or crystallised lightning. Such descriptions epitomised the unique properties of the diamond, the result of the processes that moulded its very creation.

Because a flawless diamond is composed of a single element, carbon, it is the purest of gemstones. It is also the hardest, in fact the hardest natural substance known to man.

Only a diamond can be used to scratch another diamond. And the measure of a diamond's hardness is another diamond. But though not fragile, a diamond can be split along planes; the word "cleavage" as applied to diamonds indicates the directions in which it can be optimally and cleanly cut.

And what a work of art a cut diamond is. Here, on enhanced display, are the two characteristics we most associate with diamonds—brilliance and incomparable lustre, which come from the stone's unique ability to refract or bend light as well as to reflect light. Flawless diamonds are colourless; that is the unmistakable characteristic by which they are recognised; a chemically pure, perfect crystal. This is not to say that coloured diamonds do not exist—they do, in dazzling variations, because of the presence of other elements. Traces of nitrogen make the stone yellow; of boron, blue. Other diamond colours include black, pink, green, red and even violet.

Necklace end pieces, perhaps Punjab or Punjab Hills
These unusual end pieces, rubies on one side and diamonds on the other, would have supported several strands of a necklace, perhaps of fine Basra pearls.

Yet the most desired, the most valuable, stones remain those such as were mined in India a few hundred years ago, diamonds of the truest clarity and purity, stones as brilliant and limpid as drops of water with fire in the heart.

Pendant, Jaipur
Rose cut diamond pendant set in 23 karat gold with emeralds. The reverse of the pendant is studded with ruby and turquoise.

"A NOBLE SUBSTANCE SCRATCHES THAT WHICH IS NOBLE
AND THAT WHICH IS NOT;
THE DIAMOND SCRATCHES EVEN THE RUBY.
THE DIAMOND SCRATCHES ALL AND
IS NOT SCRATCHED BY ANY".

FROM THE RATNAPARIKSHA *OF BUDDHA BHATTA*

In India, where it was first mined, the Sanskrit name for the stone is transliterated as "vajra", the thunderbolt, symbolic of the hardness of the stone and of the flash of lightning so akin to the brilliant sparkle of the diamond. Early descriptions of the gemstone date back to the fourth century BC, to the *Arthashastra*, the great text of Kautilya, minister of the Emperor Chandragupta Maurya. He lists out the diamond mining areas and the trade in the stone. Clearly by the time Kautilya wrote his thesis the diamond

Tikka, North India, 19th cent.
This finely-enamelled forehead ornament has a bird motif on the obverse which is echoed on the reverse.

was not only a highly valued substance but one which had been closely studied, for he writes of the properties of an outstanding stone: big, of brilliant lustre, with symmetrical points and capable of scratching a vessel.

An endless fascination with gemstones led to the compilation of several ancient Sanskrit texts such as the *Ratnapariksha* of Buddha Bhatta and the *Brhatsamahita* of Varahamihira. The texts reflect the intense study of stones such as the diamond, which is described in great detail and classified by shape, weight, cut, sparkle and colour. Perfection lay in the natural octahedron form of the diamond, a double pyramid with a square cross-section; the finest diamonds were defined as those which were of this shape, colourless and transparent with rainbow reflections. Such stones were the prerogative of kings, part of a caste hierarchy set out in the texts.

Necklace (haar) with pendant, and earrings, Jaipur
The long necklace (opp.) has both flat table cut and uncut diamonds set in 14 karat gold and silver. Enamelled peacocks form a motif on the sections as well as the pendant; these are separately crafted and fixed on by means of gold bolts at the back. The matching earrings echo the peacock motif and have long 'kanautis' at the top to hook into the hair.

Earrings, Jaipur
Black diamonds, small white diamonds and ruby drops are combined in this striking pair of earrings.

Perhaps this was because a spotless diamond was thought to be the abode of the gods; thus, priests and rulers could own those that had "the whiteness of conch, of the lotus, of crystal". For warriors diamonds of a reddish hue were considered auspicious; merchants could possess yellow stones with the nuance of flower petals; and lower classes had to content themselves with diamonds with the "sheen of a burnished sword", that is, grey or black. Only the king could wear diamonds of any colour, for he was the lord of all the castes.

Such distinctions might sound bizarre but the fact is that in ancient India materials were invested with meanings that resonated far beyond the surface. Diamonds, markers of the imperishable, linked to eternity, were connected to the gods through their colours: green with Vishnu, white with the Lord of the Oceans, Varuna, yellow with Indra, the god of the Heavens, copper-coloured with the Storm gods, the Maruts, and so on.

Thus beyond the indisputable and inherent value of the stone, beyond its decorative effects, lay a world of meaning that enriched perceptions of the diamond and enhanced the awe in which it was held. Its talismanic properties were both protective and healing and had a cosmic radiance that transferred to the possessor. Health, happiness and glory were written in his destiny, moreover, he was guarded against diseases like epilepsy and dangers such as serpents, floods or evil spirits.

This metaphysical resonance was most evident in the Navratna, the combination of nine gems in jewellery, that was the supreme and mystical metaphor of the Universe itself. In this arrangement, where gems are the aural and coloured channels for the vibrations of the cosmos, the diamond is associated with the supernatural energies of Shukra or Venus. Since ancient times, Shukra has symbolised beauty both of the outer world as well as inner life: diamonds, with their fiery brilliance, represented luxury, romance, the refinement of arts, prosperity and happiness.

All this was possible, the texts warn us, only if the diamond were one with "inward lustre, transparency, illumination with rays, sparkle, free from impurities and good formation of the shape". The quality of the

Earrings, Jaipur
A striking combination of black and white diamonds in an elegant pair of earrings.

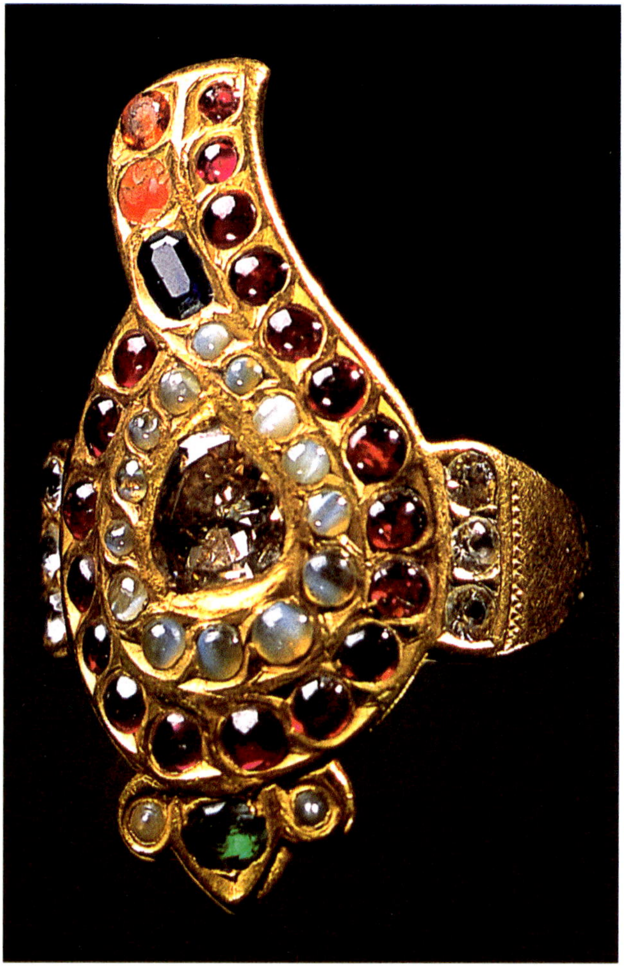

stone determined whether it was the positive or negative energies of the planet that were transmitted to the wearer; a flawed stone worked in reverse, and could bring only grief and anxiety.

Through the centuries, then, the diamond was prized in India for more than its own considerable value. It was part of mythology, philosophy and the art of healing; it represented the powerful aura of the cosmos and its profound connection with mankind; it played a part in the fortunes of kings and emperors; it drove a thriving commerce and economy.

Though it was the stones of Golconda which became the most famous in the world, there were other mines in India. As far back as the fourth century BC, Kautilya listed six principal source areas, later texts list eight. It is difficult to match the areas mentioned with present day locations. The Sambalpur mine area in Orissa stretched between the Mahanadi and Brahmani rivers; indeed, the Mahanadi has been identified as the "river of diamonds" described by Ptolemy, testifying to its antiquity as a diamond source. The northernmost group of mines was located around Panna in the Bundelkhand area of Madhya Pradesh. Wairagarh in Maharashtra was another source from remote times; perhaps it was the Vajra Grah, or abode of diamonds, mentioned in early texts.

"Jazz" ring, Jaipur
A black diamond centres this intriguingly coiled ring of white gold studded with small diamonds.

What then happened to this treasury of diamonds? By the eighteenth century, many of the mines were played out and many rendered unviable because of bad management or inefficient technology or oppressive administration, sometimes a combination of all three. The body blow came with the discovery of diamonds in Brazil, new, rich deposits; and the glory of the Indian mines passed into history. Subsequent discoveries of prime areas in South Africa only sounded the death-knell; there could be no competition between the exhaustion of the old mines and wealth of prodigious new mines such as Kimberley.

opp. Navratna man's ring, Tamil Nadu, late 19th cent.
The ring has the auspicious shape of a conch. Usually, we would find a ruby at the centre; here it is a diamond. Perhaps this was a luckier stone for the individual who commissioned the ring.

As science and technology gave a better understanding of the source and how to work it, diamond exploration expanded globally. Today, about twenty-five countries mine diamonds, South Africa of course and also Russia, Australia, Canada and several African countries amongst them Angola, Botswana and Zaire. It has been estimated that enough diamonds are mined today to give each man, woman and child in the United States a cupful! But this does not take into account the quality so crucial to fine jewellery crafting, for which only around 20 per cent of the output is suitable—the rest is put to industrial use.

What makes a diamond a prized jewel is measured by critical parameters used world-wide and known as the four Cs, that is, cut, clarity, colour and carat. The stone's rarity and hence value depends on the interplay between all four, each equally important.

Rings, Jaipur
The rings pictured here have coloured diamond centres: brown (top left), and the palest green (right).

The geometrically proportioned facets of the cut are what add fire and brilliance to the diamond. We have seen that the stone has the unique ability to reflect light; the facets enhance this play, reflecting light from one mirror-like surface to the other. With proper cutting the light enters through the top, bounces off the sides, and then travels back out of the top, giving the diamond its characteristic scintillation. However, even a slight slip in proportions can diminish the brilliance of the diamond. If the cut is too deep or shallow, light can filter out from the side or bottom of the diamond instead of flashing back to the top.

Another aspect of cut is the shape given to the stone. Cutting techniques as we know them today probably evolved from early beginnings in medieval Venice, a time when diamonds became more prominent in European royal jewellery and regalia. Early forms of fashioned shapes include the point, which followed the form of the stone's natural octahedron; thereafter, shapes (or cuts) like the table and rose. Today it is possible to have all manner of fancy shapes, but the classic cuts remain those such as the marquise, the round and the baguette. Perhaps the most popular is the round, whose symmetry and hence reflective capacity allows for the greatest

Premier Rose
Discovered in South Africa in 1978, this is the sixth largest pear-shaped diamond in the world. It has been certified as absolutely flawless with the highest qualities of colour and clarity.

De Beers diamond
De Beers' first major discovery after entering diamond mining, this light yellow diamond was once the centrepiece of the 1928 Patiala Necklace, one of the most spectacular jewels ever created.

brilliance. Such a stone could have fifty-eight facets, each placed in exact geometric relation to one another.

The greater a diamond's clarity, the more valuable it is. It is very rare to find a flawless diamond; most have minute traces of non-crystallised carbon called inclusions, often invisible to the naked eye. The fewer the inclusions the clearer the stone; in addition, the presence of inclusions interferes with the dispersion of light and hence the brilliance of the stone. Diamonds are examined very carefully under magnification and ranked on a scale ranging from F (flawless inside and out) through VVS (very very small inclusions) to SI (small inclusions). Lower grades have inclusions visible to the naked eye. Factors that determine the grade are the size, colours and numbers of inclusions.

The value of a diamond is established by its colour, rather by the absence of it, for the premium stone is one that is colourless thus permitting the brilliance of light to pass through. The singular purity of their whiteness was the special quality of the Golconda diamonds. Paradoxically, many stones may appear colourless to the eye but actually have slight traces of yellow or pale brown. The colour scale goes from D, signifying the purest, iciest white, to Z for diamonds with a strong yellow or brown tint. Colour differences are very subtle and often difficult to discern; therefore, colours are graded under controlled lighting conditions and are compared to a master set for accuracy.

It is not white or yellow alone that colours natural diamonds. There are also shades of red, pink, blue, green and black, which today have acquired a cachet and are marketed as "fancy diamonds".

Ring, Jaipur
The pear-shaped diamond in the centre of the ring is white with a pale tinge of yellow.

We measure diamonds by carats; this is actually the weight rather than the size of the stone. The word is derived from the kirat seed which was used in India to weigh diamonds (approximately 200 milligrammes) and has remained associated with the gemstone ever since. Larger diamonds are found less frequently and hence greatly prized. But size or weight of itself is no indicator.

A large stone is not very valuable if it lacks brilliance, clarity and purity of colour.

We are now back full circle to the wisdom of the Sanskrit texts. The parameters just described to evaluate a diamond are those of modern science and supported by a host of state-of-the-art laboratory devices. There can be nothing more empirical, more prosaic. But consider this uncanny echo from the *Garuda Purana* of hundreds of years ago. Without the accessories of technological advancement, this definition of a beneficent diamond is tested on the touchstone of time: "If one owns a rough diamond with a well-shaped crystalline formation, high transparency and brilliancy, and devoid of any blemish or visible inclusions, he acquires the following virtues: prosperity, longevity, marital happiness, children and livestock, and a good harvest".

opp. Golden Jubilee
The Golden Jubilee is the largest faceted diamond in the world, weighing just under 550 carats with a total of 148 perfectly symmetrical facets. The colour, graded a "fancy yellow brown", has a fiery orange depth. Its name derives from the fact that it was presented to the King of Thailand to mark the golden jubilee of his accession to the throne.

three

Treasury of the World

"...THE ACQUISITION OF SUCH RARE JEWELS...
CAN ONLY RENDER ONE SERVICE,
...THAT MAJESTY MIGHT SHINE WITH INCREASED BRILLIANCY..."

FROM THE BADSHAH NAMA, *ABDUL HAMID LAHORI*

LONG BEFORE THE BRILLIANT APOGEE of the Mughals, the vaults of rulers throughout India brimmed over with precious stones; the Mughals simply raised this accumulation to a refined art form. Apart from what came under direct control by way of mines, it was customary for kings and emperors to be fêted on auspicious occasions or during a

opp. Necklace, Ariya, Jaipur
A traditional Rajasthani design, clearly a necklace meant for the wealthy, as its name implies. A cascade of diamonds with pearls pendant flows from the collar, which has peacock motifs.

royal audience with gifts from nobles and courtiers. This nazrana, as it was called, often comprised gems and jewellery. The quality and quantity of the offering reflected the loyalty of the giver; needless to say, the value of nazranas was substantial, in direct proportion to the eagerness to find favour with the monarch. Tributes from vassals to an overlord only swelled the coffers of the king.

In addition, it was obligatory for rulers to add to the gemstone and jewellery hoards of their ancestors. Often such collections were stockpiled in hilly fortresses or buried in secret hiding places; thus the numerous accounts in Indian history of entire cities being ravaged for treasure,

Paijeb, Calcutta, early 20th cent.
The anklet is in gold and diamonds. The design shows western influences, the transition period to the Victorian style. Note the open setting of the diamonds.

of territories laid waste to uncover and secure riches. Plunder and pillage contributed greatly to a victor's spoils, be he a neighbouring raja or an invader. The historian Ferishta recorded that an early Islamic invader of India, who ravaged the country, died the possessor of close to 200 kilogrammes of diamonds, just part of the wealth he took. From medieval times to the wane of the Mughals, few conquerors departed without a sizeable booty of precious metals and stones.

We may wonder then whether this was a country of an endless dazzle of gemstones, self-renewing deposits that provided an infinity of jewels to augment the legend of India as a source of inexhaustible wealth. The thought is no mere romance. If we briefly follow the accounts of only three dynasties—those of Vijayanagara, Golconda and the Mughals—we shall see just how much this mystique is substantiated by fact and record.

Navratna bajuband, Mughal design, North India, early 19th cent.
The arrangement of gems is a metaphor for the cosmos, with each gem representing a planet. Note the carved emerald in the centre, a Mughal characteristic.

Necklace, North India
Square rose-cut diamonds and drop emeralds feature in this stunning necklace, which is enamelled on the reverse. It is inspired by jewellery from the Nizam's collection.

In its heyday, the Vijayanagara empire had a thriving capital and bazaars where diamonds were traded as casually as flowers, the roses that were "sold everywhere" and were "as necessary as food". In the sixteenth century the dynasty of the Narasimhas was more powerful than any that had ever reigned over South India. The Vijayanagara Raya had no less than sixty seaports in his empire, a vast army of a million troops and cavalry, and territories that stretched across the peninsula from west to east. His omnipotence

was described by the gold, silver and gemstones that adorned his temples, his palaces, his horses and elephants, his own person as well as those of his wives and concubines. Portuguese travellers who visited Vijayanagara could scarcely believe the enormous wealth on display around them; such unaccustomed grandeur led Domingos Paes to exclaim: "I cannot possibly describe it all, nor should I be believed if I tried to do so...I went along with my head so often turned from side to side that I was almost falling off my horse..."

Paes was allowed to see one of the imperial palaces and his awe-struck account includes a chamber "all lined with gold". This had a "four-sided porch...over which is a work of rubies and diamonds and all other kinds of precious stones...and above the porch are two pendants of gold..."

Choker, Jaipur, late 19th cent.
The velvet-backed choker is in 23 karat gold set with rubies and diamonds.

No less spectacular were the festivals and religious processions, where the dancing girls wore collars of gold enhanced with diamonds and the maids of honour, bearing golden lamps, were weighed down by jewellery so heavy that they had to be supported.

Such enormous wealth was at least partly accounted for by the rich diamond mines in the Raya's domain; at that time, they included the most productive mines in the world, such as that of Vajra Karur. In the chronicles of Fernão Nuniz there is mention of one of the Raya's top officials, "lord of the country of Gate whence come the diamonds..." Gate here referring to Gutti, a hill fort near Vajra Karur. The mines were farmed out on the condition that stones larger than twenty-five carats had to be sent to the Raya for his own use, and it would seem that there were many such.

Ear studs, Tamil Nadu
The ear studs, one of the most widespread diamond designs worn throughout the South, are in the shape of a flower head. Note the cut, faceted for sparkle.

opp. Diamond necklace
This five-strand diamond necklace is a contemporary copy of an old "sat-lada" (seven-stranded) diamond necklace. Such necklaces were seen in the royal families of Baroda and Hyderabad; this one is made of very fine quality polkis.

Choker, North India
A contemporary interpretation of an old Mughal design in gold and kundan set diamond polkis.

It is axiomatic that great riches and power are an invitation to greed and envy. The empire of Vijayanagara was surrounded by the kingdoms of the Deccan who allied against the Raya. In 1565, the battle of Talikota proved a decisive defeat for Vijayanagara and a once mighty empire was brought to the ground, its capital sacked, its palaces stripped: it took nearly six months for the victors to carry off the spoils. One of them was a diamond “as large as a hen’s egg” which was presented to Adil Shah of Bijapur. It was said that this was a gem from the plume of the head-dress of the Raya’s horse. Despite this, just about fifty years after the downfall, a king of the third Vijayanagara dynasty was said to have in his possession three large chests full of diamonds!

In the cycle of history, events replay themselves; as with Vijayanagara, so also with Golconda. The medieval Bahmani kingdom of the Deccan was a refined centre of Islamic arts and culture, a place where artists, scholars, poets and painters came together in an atmosphere of learning and exchange. In 1518 the kingdom split into five sultanates, one of which was the Qutb Shahi kingdom of Golconda, which carried on the tradition of flourishing arts backed by an informed and lavish patronage. The contribution of the diamond industry to the economy was significant, spurring all manner of trade.

The reputation of its wealth, and of its legendary mines that yielded large diamonds of the purest water, attracted the covetous attention of the Mughals, supreme connoisseurs of all that was luxurious. They wanted the diamond mines to augment their own substantial collection of precious stones and jewellery, and spared neither time nor effort to secure the conquest.

Bajuband, most probably Hyderabad, mid-19th cent.
The armlet has green enamelling set with diamonds with red and white enamelling on the reverse.

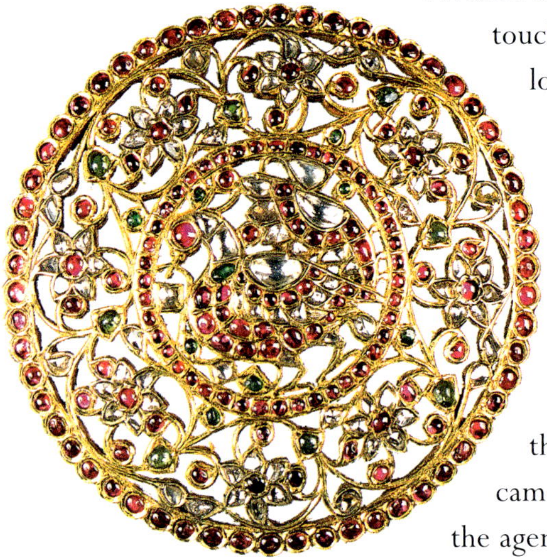

Hair ornament, Tamil Nadu, 19th cent. Part of a temple or bridal set for the hair. Here, rubies and diamonds are set in gold with a swan motif in the centre. We can see that there are many more facets in these diamonds than would be the case in northern India.

To the Mughals, gems and jewellery were glittering symbols of power and wealth, the backbone of an empire. But they were also masters of the art of aesthetic enjoyment who revelled in the creation of beauty in whatever they touched. Building on traditions that had existed for long in the country of their adoption, they added the delicacy of Persian refinement and changed the taste of Hindustan forever. They were avid collectors; from rare textiles to folios and paintings, from elaborately decorated objects in precious materials to blue-blooded animals, nothing escaped their attention or indeed their patronage. And that included the gems that poured into their treasury, whether these came from conquest or courtiers, a lapsed estate or the agents commissioned to purchase the best for their Imperial masters. At the height of the Empire, the Mughals had what has been characterised as "perhaps the greatest jewel treasury of all times".

It was Akbar, ever a meticulous recorder, who organised his gem treasury under trusted officials, classifying and valuing the considerable hoard. His son Jahangir was an unabashed lover of gems, particularly diamonds and jades. He was acknowledged as the "greatest and richest master of precious stones that inhabits the whole earth" and so vast was his collection that it was divided into 360 parts so that the Emperor could see a certain number each day. To this inspection Jahangir brought a jeweller's eye, for he was very knowledgeable about gems, unsurprising for a man whose diamond store alone was close to 190,000 carats, of which not a single stone was less than two and a half carats. His love of gems was inherited by his son, Shah Jahan, who amassed such wealth that it was estimated that it would take an expert fourteen years to go through his personal jewels. This lavishness outdid even his father's; perhaps nothing exemplifies this better than the fact that in Shah Jahan's era, there were no less than seven treasure forts in the scattered within the Mughal Empire, of which one alone, in Agra, had "tubs of uncut diamonds".

Turban crest, Rajasthan, 19th cent.
The design and precious gems used—diamonds, emeralds, rubies, pearls—indicate this must have belonged to a royal family. All sizes of diamonds have been used here, note the small flakes in the crescent. The image is from Nathdwara.

Shah Jahan was so enamoured of gems—and perhaps pressed for revenue—that it was only a matter of time before he cast his eyes towards Golconda, the prime source

Nag choti, Tamil Nadu, late 19th cent.
Worn at the back of the head, the elaborate jewel has rubies, emeralds and pearls, also diamonds. Three diamonds are flat cut, the others are faceted; and the design incorporates a Shiv linga.

opp. Pendant, 20th cent.
This is a western-inspired design, as we can see from the claw setting of the diamonds and the added faceting. The centre stone is a spinel, which was popular in the Mughal period.

of diamonds, as he was "extremely eager to possess the diamond mines". It was a long and costly campaign, carried out in stages by his son Aurangzeb, and it took more than half a century to complete the conquest. First, Golconda became a tributary state whose annual tribute was exacted in diamonds, and then fell totally to the Mughals, its royal line disinherited.

It is an irony of history that the vanquished, in a sense, outlasted the victor. It did not take long after Aurangzeb's death for the Mughal Empire to weaken "under the weight of its own contradictions", as one commentator put it. With the invasion of the Persian Nadir Shah, the remaining wealth of the Mughals was ruthlessly stripped away and carried off: one reason why so many of the Golconda diamonds now lie in vaults in Iran.

But when the centre cannot hold, new powers emerge. In the first quarter of the eighteenth century, a disaffected noble, Nizam al-Mulk Asaf Jah, set himself up as ruler without an open declaration of independence from the

Mughals. Thus was born the Asaf Jah dynasty of the Nizams of Hyderabad, accumulators of riches and jewellery of equally fabulous proportions. They inherited the Mughal mantle as collectors of gems. How vast that collection must have been is borne out by the jewellery purchased from them by the Government of India in 1995. The jewellery is only a part of the former Nizams' wealth, yet has over 25,000 diamonds weighing more than 12,000 carats!

The Nizam's jewellery, however, represents only a minute fraction of the entire jewellery tradition of India that stretches in a magnificent unbroken arc for thousands of years, from the Indus Valley civilisation to contemporary times. Like a leit-motif, one thread draws this tradition together—the superb artistry of the Indian craftsman, evident even today. Regions and epochs had their stylistic variations and favoured materials, their metaphysical themes and aspirations to beauty; all were created with precision and served by the skills of the lapidary, the goldsmith, the enamellist.

No part of the body was left undecorated; from top to toe, both men and women adorned themselves with a dazzling array of jewellery. A premium gemstone like the diamond was prominently featured in an array of necklaces, from torques and chokers to collars, chains and multiple strands; or combined with other stones in the auspicious Navratna arrangement; or set in burnished gold. There was a profusion of jewellery for the ears, such as the flower-shaped studs of the karnaphool or graceful hanging jhumkas. From upper arm to wrist, armlets, bangles, bracelets and wristlets lay next to the skin. Bejewelled bands and ornamental buckles clasped waist and hip, while the face was adorned with nose rings and the forehead with a tikka. Many a ruler had a selection of turban ornaments such as the kalgi or plume, the sarpatti that stretched across the forehead or the sarpech mounted in front. Each piece of jewellery had its own special name to identify design.

Like two great rivers that emerge from a single source, the jewellery traditions of north and south varied, though there was a give-and-take of vocabulary. What is interesting is to see how gems, and specially the diamond, were cut and set in jewellery.

Man's ring, Tamil Nadu, 20th cent.
The enamelled edges show Deccani/ Hyderabadi workmanship.

In North India, it was the tradition that diamonds were very lightly cut, if at all. The concern was for preserving as much of the original size and weight of the stone as possible, and any shaping was carried out with small facets. Stones were specially valued when the natural formation of the crystal was so perfect and clear that it was necessary only to lightly facet and polish them. Not all stones came thus; diamonds sometimes came as small chips and slices because of the mining process, and even these were kept for use in jewellery.

Stones were cut and set in different ways. As polkis, they were flat on one side and faceted to a point on the other, and usually set with the tapered side up, in the form of a rose cut. As parabs they usually had a flat table top and large numbers of such diamonds were set in traditional

opp. Man's ring, Deccan or Maharashtra, 20th cent.
Such rings were dye made and had lac fillings, but the appearance is solid.

Sarpech, Rajasthan.
Not the real thing! The stones are cut like diamonds and kundan set, but are actually Belgian glass. The turban ornament may have been created as a model, or made for someone who could not afford diamonds.

Mughal style jewellery, lightly faceted around the edges for brilliance. Flaws were hidden by covering the surface of the diamond with facets. It is believed that in eighteenth century North India, Lucknow and Benaras were the two most important cutting centres, especially for the flat and rose cut diamonds.

It was not just the cut, however, that showed the jeweller's art. Much came from the setting, where craft and skill displayed the diamonds to greatest advantage. It was the Mughals again whose refined aesthetics raised diamond jewellery to a high art; and nowhere is this better seen than in the kundan technique for setting stones. In the finest examples it brings together precious metal, gems and the delicacy of enamel art, and is a part of the jewellery repertoire much prized even today. Narrow, hammered sheets of the purest gold are literally moulded around the diamonds to encase them and hold them in place. This gives the jeweller the freedom to use stones of irregular shapes and sizes, allowing for minimal cutting. Because the stones are encased, the light strikes them only from above; so, to provide depth and refraction, a piece of foil—gold, silver or coloured—is placed under the stone to give it a

Pair of wristlets, Jaipur, 19th cent. This delicate pair has diamonds encased in the kundan technique. On the reverse (not seen) is red enamelling, hallmark of Jaipur.

Necklace, Tamil Nadu, mid to late 19th cent.
The elaborate fan-shaped design has a combination of diamonds, emeralds and rubies. Because of the cut, we can identify these diamonds as being older than the time they were re-set; and the use of the close-setting is typically Tamilian.

glow. The jewel is enamelled on the reverse with layers of finely ground enamel powders applied to engraved grooves, a process which itself demands a number of specialist skills.

There was a time when the origin of jewels could be identified through the enamelled colours. Rich red was the trademark of Jaipur, for instance; and pink that of Benaras. But the demands of royal patrons carried the craftsmen from one place to another, thus spreading both skills and forms. Mughal designs and motifs travelled to the Deccan, where they became the staples of style after the seventeenth century.

In South India, stones were set in the style called pachchikam, where diamonds were foil-backed and encased in silver much like kundan, with grooves carved from the encasing edges to make it look like open claw setting. But in traditional South Indian jewellery, diamonds were most often close-set and used with brilliantly coloured stones such as the ruby; this, together with the more obvious use of gold, gave the ornaments a rich, warm glow. A deeper symbolism infused the imaginative range of the jeweller's art in Tamil Nadu; images of the gods, of fertility, of auspiciousness, abound. This is clearly seen in the sets of hair ornaments, the long necklaces, the manga-malai or mango motif necklace, and the earrings of the temple or bridal repertoire, most often wrought in gold, rubies and diamonds. The complete set of hair ornaments or jada sringar comprised the jadainagam, which covered the braid or plait; the nagchoti worn on the crown of the head; the pendanted chuttipattam across the forehead, and the suryakala and chandrakala

Head ornaments, Tamil Nadu, 19th cent.
Such elaborately worked pieces were worn on either side of the "mang" or parting of the hair, and were essential in temple and bridal jewellery. Also called "sun" and "moon" , the design typically combines rubies and diamonds set in gold. The diamonds here are table cut.

(sun and moon), two pieces worn on either side of the centre parting of the hair. The significance of these ornaments went beyond the elaborately wrought surface. The plait had a snake-like form, and the snake was a symbol of fertility. Ornaments like this and the nagchoti were thought to harness the powerful energies held within the chakra located at the top of the head, while the "sun and moon" invoked the beneficence of the heavenly bodies.

In the late nineteenth and early twentieth centuries, as Western influences became more pronounced, Indian jewellers adapted settings to incorporate faceted stones, lighter mounts and the open or claw setting for diamonds. With this came a change in design as tastes changed and the urge for new idioms within the repertoire increased. Every new wave created its own ripples, the Victorian with its pretty florals and delicate settings, the exuberant and exotic expression of Art Nouveau with its emphasis on Western-style enamelling, the bold lines and abstract angularity of Art Deco, which cut diamonds into shapes

Necklace and Earrings, Jaipur
Flat cut diamonds find a counterpoint in tiny studded stones. The design has elements of both the local and Victorian styles.
The matching earring (opp.), one of a pair, echoes the same delicate combination.

never seen before, such as the baguette. Jewellers and their clients alike embraced the new forms with enthusiasm, adding to them their own deftness of touch; many of these expressions, adapted to contemporary taste, continue to hold great importance in prime jewellery areas like Jaipur, well-known for the skill of its craftsmen. Thus, themes and

motifs born of another culture have been absorbed and assimilated to reflect a new and very Indian sophistication in diamond jewellery.

But in the end, when we talk of Treasury of the World, we may well ask ourselves if this still applies at a time when our most prized diamonds are outside India and the greatest Indian collections remain inaccessible for one reason or another. There is reason to hope. At this stage, we should also look to the new opportunities offered by the globalisation of the jewellery trade, an opportunity we will see in the next chapters.

Most of all, we should remind ourselves of the vibrant tradition, of the unsung heroes of metal and stone and pigment, the many inherited skills, that have kept jewellery in India alive for five thousand years. And keep it alive still, despite the imperatives towards mass production.

opp. Necklace and earrings, Jaipur
The floral design is like that of the traditional karnaphool earrings.

Earring, Jaipur
One of the pair matched with the ariya necklace on page 48.

four

Ring of Fire

"...Diamonds cut and polished in India encircle the globe like a ring of fire..."

Anon

In the twentieth century, long after the diamond mines of India faded away, a phenomenon occurred. In slow stages, merchants from India revived an active interest in diamonds, this time in the cutting and polishing of the gemstones. It was a global scenario they now entered, a far cry from the old pipelines of the Deccan and transactions through secret signals; it was an arena of established players dominated by Hasidic Jews, where it was difficult to secure

opp. Nath, Rajasthan/Gujarat, 20th cent.
Nose ornament with diamond beads and set diamond with emerald. The bead cutting was probably done in Surat or Bombay.

even a toe-hold. At first they picked up the fragments that nobody else wanted, bits and chips in bare fractions of carats which were processed primarily for industrial use. Indian merchants had the advantage of low labour costs at home, so the diamonds travelled to India to be processed and re-exported. It was a foot-in-the-door approach that paid dividends; within the space of a few decades, Indian diamantaires had developed the necessary networks and skills to catapult to a position of predominance. Startling as it might sound, the fact is that today eleven out of twelve diamonds worldwide have been cut and polished in India; and the overall diamond range encompasses everything from the tiny bits used in drills and machining tools to large ten carat gemstones.

It is a remarkable story which centres around Gujarati merchants, mainly from Palanpur and Surat, whose drive, entrepreneurial skills and unique family management style brought them to the forefront of the business. From small origins, the industry has established itself as the world's largest manufacturing centre of cut and polished diamonds, contributing 60 per cent of the world's supply in terms of value, 85 per cent in terms of caratage and 92 per cent in

opp. A cornucopia of rough/uncut/unpolished diamonds.

terms of pieces. Our story of diamonds in India would be incomplete without some account of this remarkable development.

The tentative beginnings were in the years before the Second World War, when the pioneers set up shop in Amsterdam and Antwerp, the two cities that were then the main trading posts. But the great "diamond rush" really began in the years after the War with the coming of many more merchants to Antwerp. The biggest boom came in the 1980s, when the supplies of roughs were augmented by new mines in Australia and Botswana, and the insatiable Japanese drove up the demand.

Today, it is Antwerp that predominates: it is the "diamond city" through which flows 80 per cent of all the rough diamonds mined in the world. From its principal hubs in Antwerp and India, the Gujarati diamond network spans the globe, from Hong Kong and mainland China to Bangkok, the Middle East, Europe and the United States. The context is truly international, with both suppliers and buyers from many different countries. Within India, over one million workers—mainly in and around Surat and Mumbai—carry out the task of perfecting the stones for use.

Centenary
The discovery of "one of the largest top colour diamonds ever found" was announced in 1988, when De Beers was celebrating its centenary, hence the name. When cutting was completed, the Centenary weighed 273.85 carats and had 247 facets—164 on the stone and 83 around its girdle, bringing out the fire and brilliance of the stone.

Bracelet, Jaipur
Western-style bracelet, cut and uncut diamonds set in 14 karat gold and oxidised silver form a delicate floral motif.

For all that it is a global network, it is still a closed one which operates within the security of family groups and where business is conducted with only known partners worthy of confidence and trust. It is a business where much remains unwritten in the full knowledge that a word will be honoured, and where diamonds valued at crores of rupees are hand-carried from one place to another by the traditional Gujarati courier service called angadia. No written records or receipts are exchanged, but the angadias have a successful delivery rate of 99 per cent.

Because the families prefer to keep a low profile, the value of their business is not always fully appreciated. But statistics tell the story more dramatically. Gems and jewellery contribute significantly to India's export trade (about 17 per cent, second only to software and outsourcing); of this, diamonds alone account for 80 per cent. What is more, the export of cut and polished diamonds has had an almost exponential growth, rising by 28 per cent in a single year, and this at a time when more

opp. Collar and earrings, Jaipur
Round and baguette cut diamonds sparkle in a setting of 18 karat gold. The design is clearly western-inspired, following a kind of Art Deco style.

and more stones are being used in jewellery sold domestically. Side by side, the export of diamond jewellery has seen a phenomenal rise, 78 per cent in one year.

Indeed, if there is any factor that is persuading the big families to come out of the shadows, it is their forays into diamond jewellery, a natural extension of the value-added chain. As yet, though, the core business remains cutting and polishing.

There is actually an irony when we consider these activities. In the olden days of Indian diamonds, the art of cutting was practiced in an altogether different way. Then, as we have seen, the main concern was to preserve as much of the stone size and weight as possible, thus cutting was minimal. Facets were made more to shape stones, or mask flaws, than to uncover brilliance. Today it is a completely different story.

And it is different in more ways than one. We are fed on the romance of diamond cutting, on stories of venerable master cutters

who study stones for days, months, perhaps even years, before essaying the cut. So fraught was the cut of the massive Cullinan diamond, at 3,106 carats the largest diamond crystal ever found, that a doctor was brought in to look after the master cutter: certainly medical assistance was required, as he fainted after the second and successful try. Models have been made of large diamonds so that stones can be studied and analysed from many perspectives, including light images, for the perfect cut. In the late twentieth century, master cutter Gabi Tolkowsky took close to three years to complete the cut of the Centenary, about 600 carats in rough form, which he fashioned into the world's largest modern-cut flawless diamond.

The Indian reality of today does not follow any of these scenarios. For a start, most of the stones are much, much smaller than the ones just mentioned. Diamond cutting and polishing, like any other industrial process, follows an established chain of events in an almost assembly line fashion, and the pipelines stretch across the globe, usually from London and Europe through India and back again to

opp.
top: Manual inspection of a rough through a 10xLoupe.
below: Inspection of diamond purity under magnification.

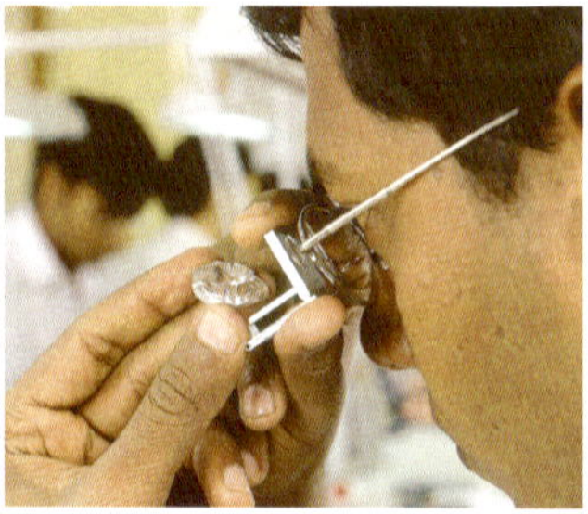

destinations around the world, including the United States, Hong Kong, Japan and Thailand. London is the headquarters for the Diamond Trading Company (DTC), part of the giant De Beers Group which sorts rough diamonds from pretty much all over the world, including their own mines in Africa, into thousands of categories of shapes, sizes, quality and colour. The stones are evaluated and then sold to a select list of established clients known as sightholders, who meet in London ten times a year to "sight" their allocations, for which they must pay in advance. It is a telling fact that the number of Indian sightholders has increased over the years until today they form more than one-third of the list and absorb just under a quarter of total DTC sales.

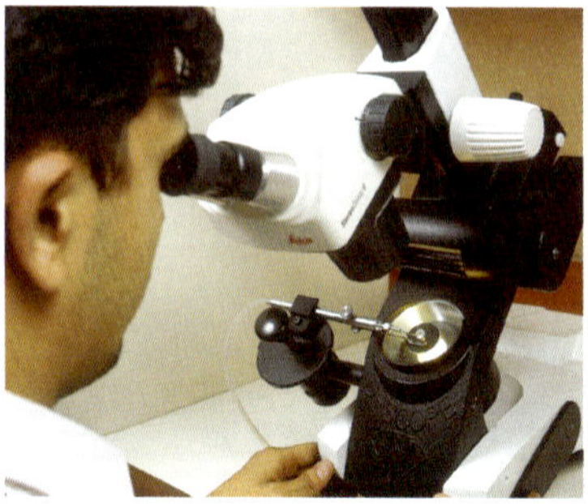

A sightholder like Blue Star, for example, symbolises the success story of Indian diamonds. A group which spans three continents and employs over 3,000 people worldwide, it is one of the largest diamond manufacturers in dollar terms and has its state-of-the-art flagship factory at Varachha near Surat.

Bruting-shaping of a rough diamond

At Blue Star, when the roughs are received they are placed in packets for easy tracing and sent to Surat. Here, planners inspect the stones to see how best to maximise the beauty and value of each. A small window polished on the rough crystal allows them to see inclusions and determine the shape and size. Once this is done, the experienced eyes of assorters make sure that all is in order.

Skilled operators now take over the process, using lathes and lasers at various stages to brute and block the stones, shaping them first into pyramidal forms and then creating the final cuts. The stone is girdled for delicate edges and then faceted to refract the fiery light we associate with diamonds. Throughout, the concern is to preserve as much of the value of the stone as possible with special attention to its inherent purity, colour and size. At the end of the process, a perfectly shaped and sparkling diamond with the traditional fifty-eight facets emerges, ready to be exported for use in jewellery.

Defining the shape of the rough

Blue Star is one of the many companies that is, literally, shaping the diamond business. Another is Rosyblue, today the world's largest diamond manufacturer. The heady

growth experienced by such companies carries with it more challenges—but also new opportunities. One of the biggest now facing the established players is increasing the size and value of the diamonds they cut. Traditionally, it was only smaller size stones that came to India to be cut, but the fact is that the very high value stones, numbering only around 10 per cent, account for half the value of the world's diamonds. This is the apex of the diamond league; such stones are still cut in places like Antwerp, New York and Tel Aviv. A venture into this top end can only be made when the skills are to hand; those skills are gradually being taught to Indian diamond cutters by masters in Israel and Belgium. Perhaps it will not be too long before significant inroads are made into this and other lucrative segments too.

Round cut brilliant and pear shaped diamonds.

Certainly it will not be for lack of trying. The Bharat Diamond Bourse, the world's largest, will be operational soon in the city of Mumbai, and aims to process 90 per cent of India's diamond trade. Spread over twenty acres, the complex will provide world-class facilities to the members of the diamond community and enhance the already predominant position of the Indian diamond

Necklace, Hansuli, Jaipur
This sleek Art Deco style presentation, seen in close-up, combines Belgian baguette and rose cut diamonds shimmering in gold.

industry in the world market. Planned as a single-window operation, it will have its own dedicated customhouse facility and international standard security measures to serve the interests of small and large diamond traders.

Despite these thrusts, there are still factors that are of concern. India does not, as yet, produce diamonds, thus the question looms large of ensuring a smooth and steady supply of roughs for processing. The supply pipeline is controlled; moreover, a mine is not an infinite source of stones, and has a productive life which sooner or later comes to an end. There is another twist to this tale. Whereas cutting centres like India are trying to source directly from mining countries, they in turn are trying to promote cutting on their own soil, bringing the manufacturing process within their ambit. This may well be a logical extension for them, but poses a risk to Indian entrepreneurs and their hundreds of thousands of skilled workers. As a responsive strategy many members of the Indian diamond industry have gone into jewellery manufacture for both export and domestic markets, taking activity right upto the retail level. Equally, new ventures are looking to prospect for diamonds in Madhya Pradesh, Jharkhand and around the Kolar and Hutti gold fields. And who knows, it might well be that the glory days of indigenous diamonds are revived...

A larger issue confronting the entire diamond world is that of the rapidly proliferating treated or synthetic manufactures of the stones themselves. There are already a number of these: treated stones, for example, can alter colour or clarity through processes such as High Pressure, High Temperature (HPHT) and Laser Drilling respectively. Synthetics are laboratory grown products, crystallised through artificial intervention such as chemical vapour deposition (CVD). There are also simulants, natural or manmade stones that are cut just the way a diamond is. Glass is the oldest example; Cubic Zirconium the most widely available, and Moissanite the most recent entrant. Synthetics can be detected through

machinery and testing, but some have been sold as the real thing. Legally, there is nothing to stop the manufacture of synthetics as long as they are declared as such and not used fraudulently. Indeed, perhaps—like cultured pearls—it is an idea whose time has come, though it must be said here that a German court has ruled that the word "cultured" cannot be applied to synthetic stones.

The question remains, however, about the impact that mass manufacture of such stones will have on natural diamonds, not just in the marketplace but also in the consumer's mind. Research, in fact, has shown that as far as diamond jewellery goes, people want the real thing. Simply put, its perceived value derives from its rarity and antiquity, a stone billions of years old that the earth isn't making any more.

Whatever the future holds—and it is viewed with optimism by many—the rise and growth of the diamond industry in India has been a remarkable feat. Not just because of the drive and entrepreneurship of the Gujarati merchants but also because of the enormous skills pool they have helped create; assets which include diamond factories which rank amongst the world's best, using cutting-edge technology and the adaptive genius of the Indian artisan.

opp. Necklace, Rajasthan
Not diamonds but topaz. But note how the foil backing and the kundan setting give the appearance of diamonds.

five

The Last Frontier:

BRANDED DIAMOND JEWELLERY

"The emerging Indian woman has started using tradition instead of being used by it.
The opportunity for contemporising heritage jewellery forms has never been better."

C.K. Venkatraman, COO Tanishq

Jewellery in India has been called "one of the last great commodity frontiers"; its unorganised nature makes it an exciting, yet fragmented, area. The total market is valued at Rs. 65,000 crores, of which diamond jewellery constitutes just about one-tenth at a figure estimated over

opp. Necklace: Yatra, from Tanishq's Design Studio (Sumit Jain), one of the top three worldwide entries at the "Diamonds: Nature's Miracles" competition, 2004.
A course of uneven web-like strands gradually transforms into a constellation of sparkling brilliance.

Rs. 6,000 crores. The latter is a market that is growing by leaps and bounds, one that is ripe for the attention of the branding exercise and all that it entails.

But in India, think jewellery and you think gold. Gold is our statement, our investment, our safety net, liquid, imperishable. It has been the choice of millions and for centuries; close to one third of the world's gold is held in India, much of it as jewellery in private hands; and to this day, we remain the largest consumers and fabricators of the precious metal.

Bracelet, from the Mughal Collection.

Think jewellery, and you think of the family jeweller, a small or large concern with whom a relationship has developed, perhaps over generations, and who knows your every jewel because he has made it or re-set it, maybe even more than once. He knows your taste, he is mindful of your budget, he has served more than one family member: avuncular or brotherly, the comfort and familiarity levels are high, and so is confidence.

It is not an easy job to take on such core beliefs and alter perceptions. On the other hand, we live in a society where established perceptions are challenged on a daily basis through the manthan, the churning, of globalisation and the lightning speed of communications, the rise of middle class incomes and working women, the creation of icons and celebrities, a media-heightened awareness of aspirations and, what is important, the willingness and means to satisfy them.

Valentine ring, with a combination of inset diamonds and semi-precious stones, symbolises the fusion of souls, of yin and yang.

It is in this social context that we may choose to see the entry of branded diamond jewellery. Or we may view it as an outcome of other factors. For Tanishq it was the logical extension to success in watches. For Orra-The Diamond Destination, it represented the last link in the value chain that went "from rough to retail", the parent company, Rosyblue, being the world's largest diamond manufacturer. Rosyblue is one of many manufacturers who see growth in this diversification of both products and markets; and potential changes in the global diamond industry simply highlight the sense of such a strategy. Thus, in the last half-decade, the number of players in the branded diamond jewellery market has shot up.

But perhaps the biggest thrust has come from the Diamond Trading Company, the marketing arm of the giant De Beers Group which controls access to over half the world's rough diamonds. Seeing the enormous opportunity offered by the Indian market, expanding at double digit growth rates, both context and timing were right for branding. Today, branded jewellery accounts for between 8 per cent and 10 per cent of the total diamond jewellery market, and given the lavish promotion and high-profile tie-ups which ensure attention, this figure can only grow. The DTC's brief is to achieve this growth by opening new consumer bases to the awareness of diamonds and diamond jewellery.

In practical terms, this translates into stimulating the creation of more brands and supporting these with advertising and distribution. Strong brands, DTC believes, are an essential ingredient in the transformation of the market for diamonds. And multiple competing brands create the sort of excitement which brings in new consumers. DTC's role is that of brand creator,

Solitaire earrings from the premium Arisia Collection (top) and (right) earrings from the Asmi Collection, symbolic of the inner fire and identity of the modern Indian woman.

which it does by having a clear brand concept which is communicated to sightholders who then design and manufacture the jewellery to be sold through select retailers. Nakshatra, the first of the DTC launches, achieved an almost iconic status through its powerful design concept of seven-stone florals; whereas Sangini, aimed for the spouse gift market, was backed by the concept of the wife being a very special person to her husband. At the same time, DTC works with a number of jewellers including Tanishq and Orra-The Diamond Destination, who have their independent diamond collections, by lending promotional or marketing support.

Bangles and a ring: a delicate and aesthetic use of metal and stone.

It is this strong marketing and sponsorship support that has brought branded diamond jewellery so much to the forefront. And what has that done to gold? With the traditional values attached to it, plain gold jewellery is seen as slightly old fashioned, not quite contemporary: "gold carries the

burden of the past," as Orra's Vijay Jain puts it. Whereas diamonds are not just forever, they are for everybody and at prices people can afford.

This last perception has changed the way people look at diamonds. In the traditional jewellery repertoire, diamonds and diamond sets were the ultimate symbol of luxury and meant for the very, very rich. Most of the time they were put away in lockers, to be brought out only when required for wearing. Branded jewellery, backed by heavy promotion, has driven home the message that this is no longer true: diamonds are not only for you and me, but they can be worn every day.

Necklace and pendant from the Sangini Collection, designed as the ultimate gift from husband to wife. The design is based on the concept of being special to the spouse; the centre stone of each piece is of slightly larger size than the other diamonds.

The message falls on fertile ground; in part this derives from how and why jewellery itself is worn today. It has long been the custom in India that a woman, and especially a

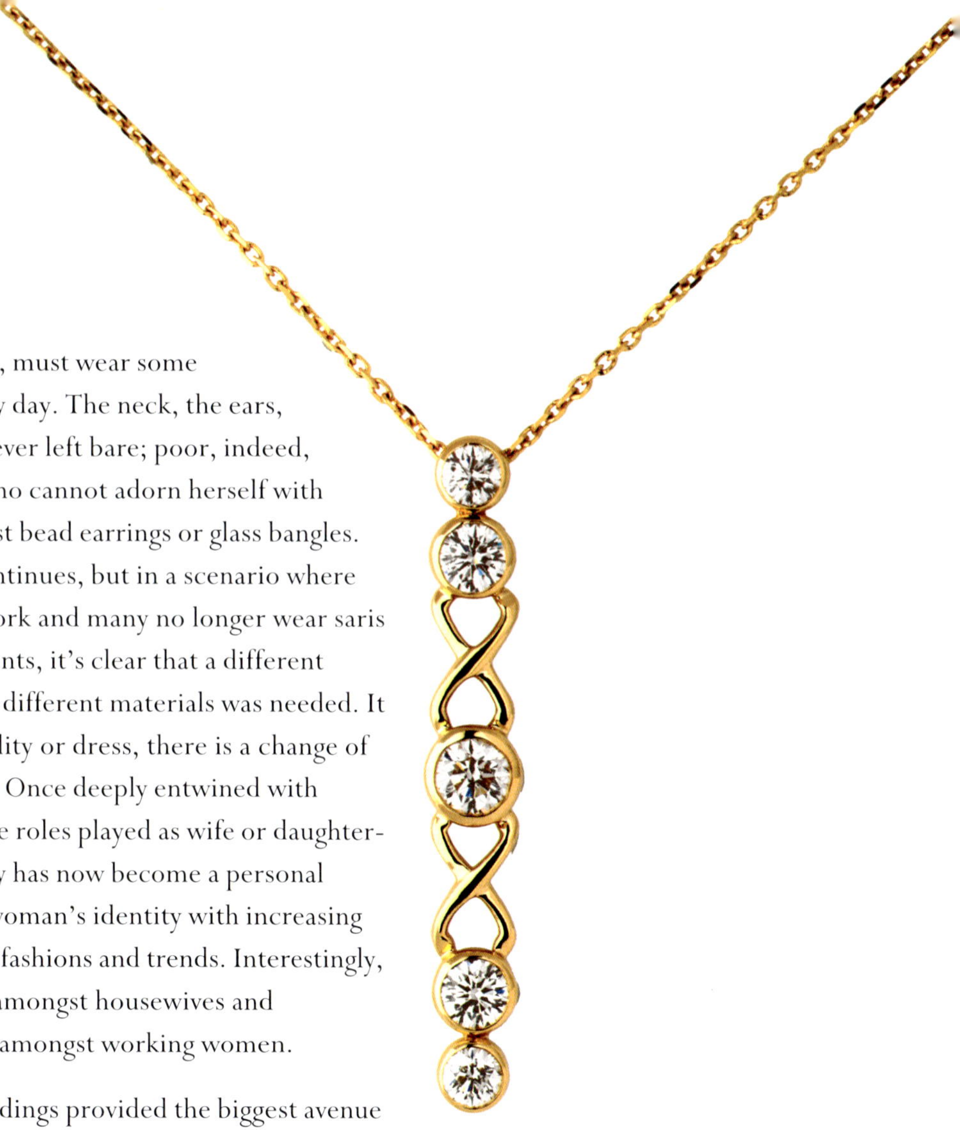

married woman, must wear some ornaments every day. The neck, the ears, the wrists are never left bare; poor, indeed, is the woman who cannot adorn herself with even the simplest bead earrings or glass bangles. This custom continues, but in a scenario where women go to work and many no longer wear saris or Indian garments, it's clear that a different vocabulary with different materials was needed. It is not just mobility or dress, there is a change of mindset as well. Once deeply entwined with tradition and the roles played as wife or daughter-in-law, jewellery has now become a personal statement of a woman's identity with increasing connotations of fashions and trends. Interestingly, this is as much amongst housewives and homemakers as amongst working women.

In the past, weddings provided the biggest avenue for purchase of jewellery, with a certain amount bought during festivals and auspicious seasons. That fact remains, but today jewellery buying occasions have multiplied. Jewels are part of a lifestyle where gifting has become big

business, so anniversaries, the birth of a child, a promotion at work, a celebration, are marked with these expressions of value. This has been spurred by the fact of rising incomes, not only amongst men but women too, who may opt to buy themselves jewellery as a matter of personal choice.

Such circumstances shaped the positioning of branded diamond jewellery. If the market was to grow, if people had to be encouraged to buy diamonds, then the ground had to be shifted from elitism to a more inclusive, more mass consumer base. A range of profiles had to be catered to, and preconceptions of age and affluence discarded. The organised sector had to do what it did best—use marketing tools to ferret out the gaps and niches that remained unfulfilled by traditional jewellers; bring into play design talents to create excitement; and put into place the distribution network so critical to accessing the consumer through multiple retail outlets. Added to this was the advantage of a totally transparent operation, where purchases could be certified and the consumer sure of what she

Pendant (top) and earring from the Nakshatra Collection. The delicate florals are a design concept which give this brand a strong identity.

was buying. These factors have contributed to the growing success of branded diamond jewellery.

Design is a vital component. As more entrants compete in the marketplace, design is a "differentiator" and theme-based or concept-based collections are launched on a regular basis by all major players. Tanishq, the pioneer in branded diamond jewellery, started a collection called Light-weight diamonds, necklets with small diamonds, where the design modification made the price within reach. Daytimes, yet another collection, widened the field even further. Its low start price was meant for working women, home makers, college girls, something they could wear every day. Orra has five global design centres, and the interchange of ideas and themes is a vibrant and lively cross-pollination. The company refreshes its design range with five to six collections a year, each with a theme that is carried into the smallest detail of execution. Their Koh-i-noor collection reflected the values of a heritage expressed in an elegant contemporary manner. For DTC, branding is about attributes; thus, Asmi ("I am" in Sanskrit) embodies the inner fire, the independence and the identity of the modern Indian woman. For designers, there is scope to

Ring, the Sangini Collection. Note the unique bezel setting of the centre diamond, which is larger than the others.

combine the old with the new, to create a fusion between Indian concepts and Western styling and metals. Diamonds studded in white gold have a sleek finish, very much in tune with the look of the times.

At the apex of the consumer pyramid are the wealthiest consumers who, in fact, own and buy the largest number of diamonds. This is the traditional domain of private and family jewellers, but here too, branded jewellery is making inroads with special offerings aimed at this segment. DTC's Arisia is a collection of the finest solitaires of one carat or more comprising rings, earrings and pendants set in gold. Orra's large stones are set in platinum, as premium as you can get; and the Tanishq Zoya collection features intricate, handcrafted designs with diamonds, rubies, emeralds and semi-precious stones set in yellow and white gold.

But however spectacular the design, however imaginative the events and promotions that accompany each launch, finally it is all about having the network of retail stores where people can buy the jewellery. Tanishq is India's largest jewellery retail store chain with a presence in fifty-four cities. And Orra is looking for fast-paced expansion; with twenty-four stores in fifteen cities already, they are planning to add at least ten each year over the next three years.

opp. Tamarine necklace, a one-of-a-kind combination of rare natural coloured diamonds in brown, cognac and white. Named after the tamarind tree, the million-dollar necklace has a warm tropical glow that evokes its exotic splendour.

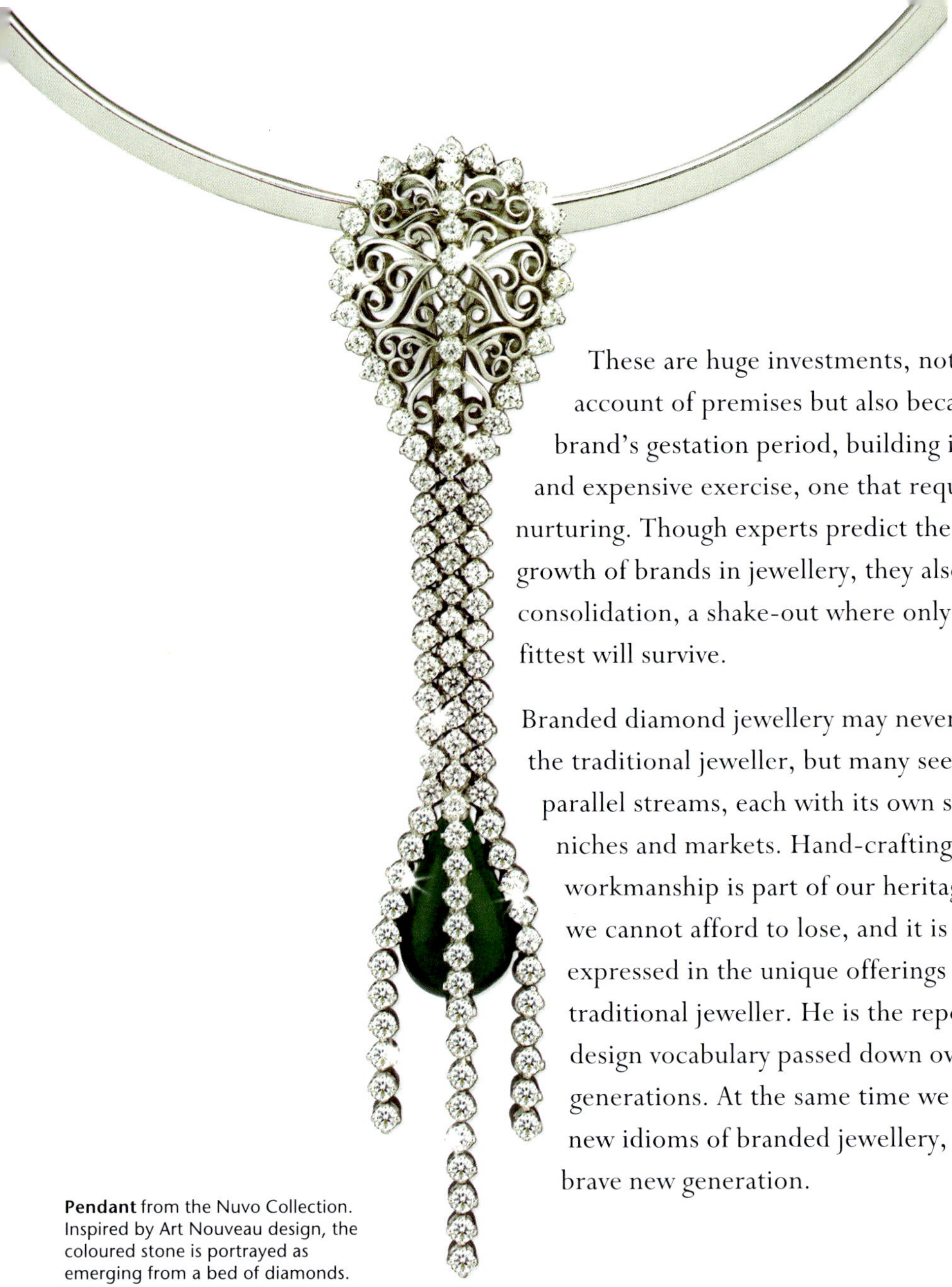

These are huge investments, not just on account of premises but also because, given a brand's gestation period, building it is a lengthy and expensive exercise, one that requires constant nurturing. Though experts predict the further growth of brands in jewellery, they also foresee a consolidation, a shake-out where only the largest or fittest will survive.

Branded diamond jewellery may never quite replace the traditional jeweller, but many see both as parallel streams, each with its own strengths, niches and markets. Hand-crafting and exquisite workmanship is part of our heritage, one that we cannot afford to lose, and it is best expressed in the unique offerings of the traditional jeweller. He is the repository of a design vocabulary passed down over generations. At the same time we need the fresh new idioms of branded jewellery, the voice of a brave new generation.

Pendant from the Nuvo Collection. Inspired by Art Nouveau design, the coloured stone is portrayed as emerging from a bed of diamonds.

Principal Bibliography

Traditional Jewellery of India
Oppi Untracht,
Harry Abrams, New York, 1997

Dance of the Peacock: Jewellery Traditions of India
Usha R. Bala Krishnan and Meera S. Kumar,
India Book House, Bombay, 1999

Romance of the Golconda Diamonds
Omar Khalidi,
Mapin Publishing, Ahmedabad, 1999

Minerals and Metals in pre-modern India
A. K. Biswas,
DK Printworld, New Delhi 2001

Diamonds:
Hubert Bari and Violaine Sautter
Vilo International, Paris 2001

Jewels of the Nizams
Usha R. Bala Krishnan
Min. of Culture, New Delhi
and India Book House, Mumbai 2001

Travels in India, Vols 1 and 2
Jean-Baptiste Tavernier, ed V. Ball
Oriental Books Reprint Corporation, New Delhi 1977

The Forgotten Empire—Vijayanagar
Robert Sewell
National Book Trust, New Delhi, 1962

Internet:

(a) The Nature of Diamonds:
Text of the American Museum of Natural History
Exhibition, 1997-98

(b) Website of the Gems and Jewellery Export Promotion
Council

Acknowledgements and Picture Credits

For permisssion to photograph from their collections, and for their assistance, we would like to thank:

The National Museum, New Delhi; Dr. Rita Devi Sharma
pp 4, 8, 10, 11, 12, 13, 16, 18, 19, 22, 24, 40, 59 to 67, 72, 86
photographer Avinash Pasricha

Bharany's, New Delhi; C. L. Bharany and Ramji Bharany
pp 34, 50, 52, 55 to 57; photographer Avinash Pasricha

Gem Palace, Jaipur; Sudhir Kasliwal
pp 2, 20, 21, 26, 31, 32, 33, 53; photographer Mahesh Hariani

Modi's The Ganpati Gems and Arts, Jaipur; Pawan and Pinku Modi
pp 3, 35 (left), 38, 41, 42, 45, 84; photographer Mahesh Hariani

Bhuramal Rajmal Surana, Jaipur; Pracheer Surana
pp 9, 36, 37, 48, 68 to 71, 78, 79; photographer Mahesh Hariani

Private collectors who graciously allowed photography
pp 6, 15, 35 (right), 39, 51, 54, 58
photographers Avinash Pasricha and Mahesh Hariani (p 39)

For permission to use images from their collections, and information and interviews, we would like to thank the following:

The Diamond Trading Company, India; The Diamond Information Centre, India, both in Mumbai; Marina Raj pp 28, 43, 44, 46, 76, 92, 94 to 97

Tanishq, Bangalore; Bhuwan Gaurav (Delhi) and Rupal Sheth (Bangalore) pp 1, 88, 100

Orra-The Diamond Destination, Mumbai; Vijay Jain and Malini Nair pp 90, 91, 93, 98

Blue Star, Mumbai; Akshay Tulshan pp 30, 75, 80 to 83